BY V.C. THOMPSON

CONSPIRACY THEORIES DEBUNKED

TIME TRAVEL IS REAL

45TH PARALLEL PRESS

Published in the United States of America by Cherry Lake Publishing

Ann Arbor, Michigan

www.cherrylakepublishing.com

Editorial Consultant: Dr. Virginia Loh-Hagan, EdD, Literacy, San Diego State University

Reading Adviser: Beth Walker Gambro, MS, Ed., Reading Consultant, Yorkville, IL

Photo Credits: © Jessica Rogner, cover, 1, interior graphics; © rad fx/Shutterstock, 4; © andrey_l/Shutterstock, 7; ©DreamcatcherDiana/Shutterstock, 9; © agsandrew/Shutterstock, 11; © Virtual Museum, 13; © Jamie/flickr/Liberty (Canadian edition), 14; © Wilson Kwong/Shutterstock, 17; © PXhere/Public Domain, 19; © The Yorck Project/Wikimedia, 21; © Gorodenkoff/Shutterstock, 22; © Luca Giarelli/Wikimedia, 25; ©UMB-O/Shutterstock, 26; © Anastasios71/Shutterstock, 29

45th Parallel Press is an imprint of Cherry Lake Publishing Group.

Library of Congress Cataloging-in-Publication Data has been filed and is available at catalog.loc.gov

ABOUT THE AUTHOR

V. C. Thompson is a nurse and author living in Michigan. V. C. is husband to his lovely wife, Anna. Together they are the proud parents of two children, one in Heaven and one newborn. His mission in life is to share his faith through the written word and the healing hand.

TABLE OF CONTENTS

There are many unsolved mysteries. The mystery of Walter Collins is one. Walter went missing. The police found him. But the boy they found wasn't actually Walter. Yet police insisted he was Walter.

INTRODUCTION

The world is filled with mysteries. Mysteries are things or events that cannot be easily explained. Some mysteries are conspiracies. Conspiracies are secret plans made by a group of people. Most conspiracies are stories that have no evidence. Evidence is proof. It is facts or information that supports a claim. Some conspiracies are later found out to be true. But most conspiracies are false. Sometimes it is hard to tell which conspiracies are true and which are false.

People who believe in conspiracies are called conspiracy theorists. Theorists are people who explain things with ideas called theories. Most people think conspiracy theorists are wrong. Some conspiracy theories are easy to prove wrong. Others are harder to prove wrong. Some are funny. Others are very serious.

Some people call conspiracy theorists crazy or stupid. This is hurtful. It is wrong to attack people, even if what they think is incorrect. Instead, the conspiracy theory should be examined. To examine a conspiracy theory, look at the evidence. Almost every claim has evidence. What matters is *how much* evidence there is and how strong it is. This tells us if the theory is true or false.

Let's take a look at a popular conspiracy theory. Don't forget to keep track of all the evidence. At the end, see if you can debunk this conspiracy yourself!

The world is fascinated with time travel. There are many books and movies about it.

CHAPTER 1:

THE PREMISE

Time travel. We've all thought about it. Who wouldn't want to see living dinosaurs? Who wouldn't want to see what the world looks like 50 years from now? But is it possible? Does the technology already exist? Some people imagine traveling using a time machine. Some think about traveling using a portal. Portals are passageways. These portals are said to be hidden throughout Earth. Some think about seeing the past or future. They do this using a crystal ball.

Time is like a line. The line began when the universe was created. There are 3 parts to the line. The first part is the past. The past line grows longer with each second that passes. The next part of the line is the present. The present is what is happening right now. Think about it as a dot in the middle. Finally, there is the last part. This part is the future. The future is the part of the line after the "present" dot. We can remember the past. We can think and move toward the future. But we can only experience the present.

What if we could move that dot backward? What if we could slide it forward? That is time travel. It sounds impossible. It sounds like science fiction. Science fiction is imaginary stories written about space, technology, or science. But there is evidence that the science is out there.

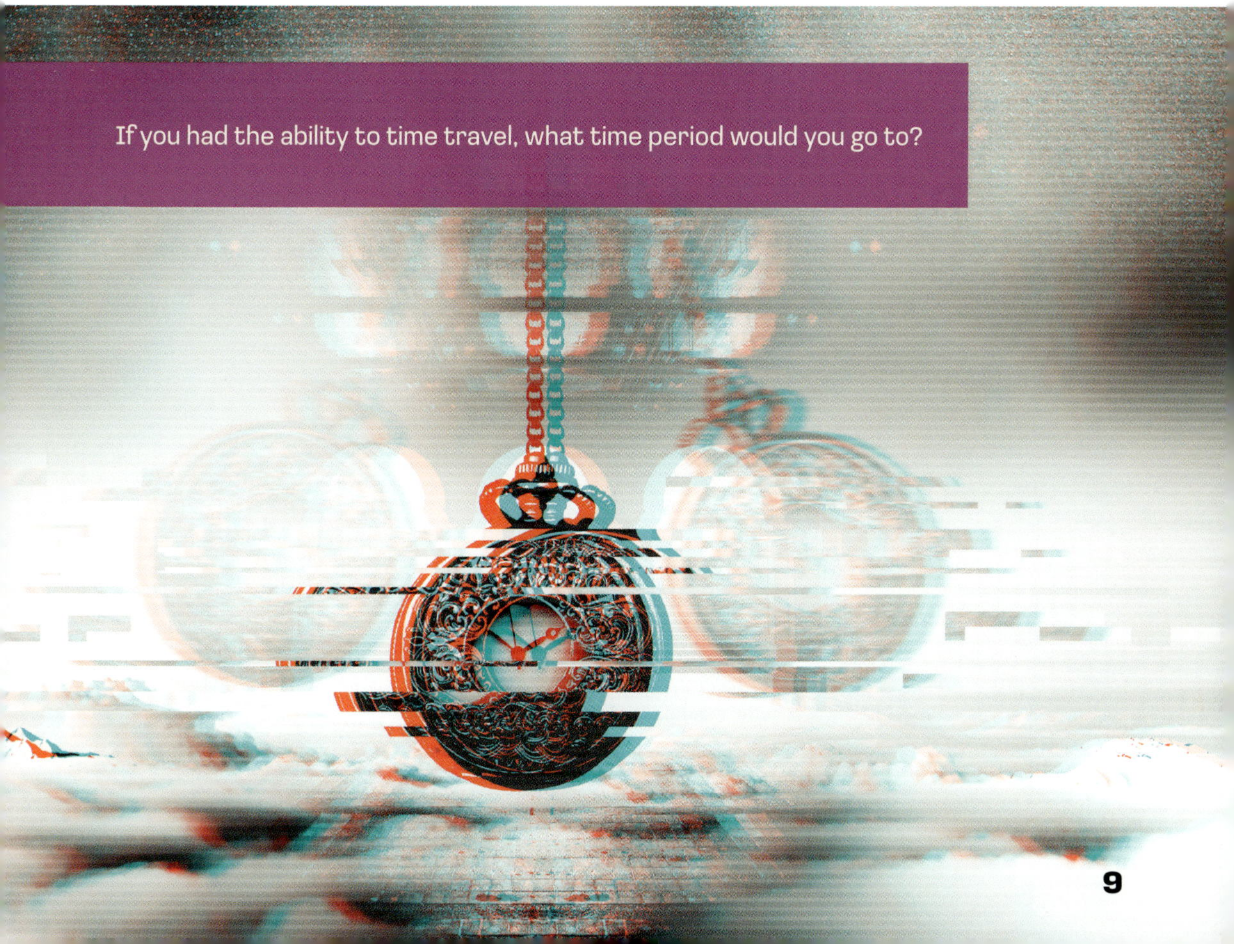

If you had the ability to time travel, what time period would you go to?

SPOT-LIGHT

ALBERT EINSTEIN

Albert Einstein was a very intelligent scientist. He was born in 1879. He died in 1955. He had an idea called the theory of special relativity. This theory states that the speed of time is relative. Relative means something that is connected to or dependent on something else. For example, light is the fastest thing in the universe. It moves at 983,571,056 feet (299,792,458 meters) every second. If we can move faster than the speed of light, we can time travel.

There was an experiment in 1971. Scientists put 4 accurate clocks on different planes. The clocks read the same time. Some planes went east and some went west. One plane stayed on the ground. The planes returned. The scientists compared the clocks. The clocks in the planes that went east were running slightly later. The clocks in the planes that went west were running slightly faster. This is called time dilation. This experiment supports Einstein's theory.

Many people believe time travel to be real. There is a great deal of evidence for this conspiracy. Let's examine it further.

CHAPTER 2:

PHOTOGRAPHS

THE CONSPIRACY

What better proof of time travel is there than pictures? Many photos of time travelers are said to exist. For example, a photo was taken during the opening of the South Fork Bridge in British Columbia, Canada. It was in 1941. The photo shows a man wearing a graphic T-shirt. He's also wearing sunglasses. This style of clothing didn't exist back then.

Another famous photo shows people on vacation in 1943. They're at a beach. Everyone is dressed in the style of that time period except for a particular man. This man is looking down at a cell phone. But cell phones were invented in the 1970s and didn't become commonplace until the 1990s . The man is also wearing a business suit! Who wears a business suit to the beach?

These people took modern technology back in time with them. They have to be time travelers. Right?

This is the 1941 photo. It shows a modern-looking man in a T-shirt and sunglasses. T-shirts became popular during the 1960s.

BE "Mouth-Happy"

Smoking used to be more popular in the past. Back then, people weren't aware of the health risks. Today, we know better.

DEBUNKED

Photos of time travelers look convincing. But this doesn't mean they are true. We are very familiar with our own time. So we look at everything through our own perspective. That means we see what we are used to seeing. Our brains sometimes project information to make sense of things. Project means to assign one's own ideas to others.

Let's look at the South Fork Bridge photo from 1941. The man looks "modern" because he's dressed casually. Everyone else in the photo is dressed in suits. But everything the man is wearing was already invented. His sunglasses were invented in the 1920s. His graphic T-shirt isn't actually a T-shirt! It's a sweater. The sweater has an emblem sewn into it. An emblem is a recognizable symbol. This emblem was the logo of a hockey team.

The beach photo is another example. The man appears to be holding a phone. But in reality, he is most likely rolling a cigarette. Many people did this back then. But few people roll cigarettes now. We aren't used to seeing this action. So we project information of what we know and fill in the gaps. It's true that wearing a business suit to the beach is unusual. But this doesn't mean he's a time traveler!

CHAPTER 3:

VIDEOS

THE CONSPIRACY

Charlie Chaplin was a famous actor in the 1920s. In 1928, he made a movie called *The Circus*. In it, a woman in the background walks across the screen. She is holding a phone to her ear. She is even seen talking into the phone. Portable phones were not invented for another 45 years!

Another case of time travel involved a man named Håkan Nordkvist. Nordkvist lived in Sweden. One day in 2006, his sink started leaking. He tried to fix it. But under the sink, he found a time portal! When he crawled through it, he was in the future. He traveled to 2046! He met his future self. He took a video using his cell phone. You can see both versions of Nordkvist hugging each other. They even have the same tattoo in the same place.

There are many time travel theories. All theories introduce a paradox. Paradoxes are contradictions.

SPOT-LIGHT

BIAS AND REASONING

A good researcher knows to dig deeper. They know that evidence needs to be examined. They need to know if the evidence is biased. Biased means favoring one idea because of preference.

Sometimes we only see what we want to see. Our experience and knowledge affect our perception. They also affect our reasoning. This is called confirmation bias and motivated reasoning. Confirmation bias is when we only notice evidence that supports our view. Motivated reasoning is when we quickly accept information that we already agree with. We examine further information that we don't agree with.

Conspiracy theorists can fall into this trap. While researching we might only look for what we want to find. We might only be open to certain evidence. But there is an easy solution. Be open to all evidence. Use credible sources. Credible means trusted.

DEBUNKED

Similar to the debunked photos, people project information to fill in the gaps. In the Charlie Chaplain movie, the woman isn't holding a phone. She is holding a hearing aid! Hearing aids back then were large and had to be held to the ear. Today's hearing aids are a lot smaller. They fit snugly in the ear. We are familiar with this type of hearing aid. We are also familiar with holding phones to our ears. That is why many people thought it was a phone.

But how can we explain the video of Håkan Nordkvist? That is easy! The video was actually a commercial for an insurance company. Both men in the ad were actors. Clips of the video went viral on TikTok. Many people didn't do further research. They thought it was real!

CHAPTER 4:

PAINTINGS

THE CONSPIRACY

Time travel has been documented in paintings. For example, *The Expected One* is a painting that shows a boy and a girl outside. The boy is holding a flower. He is waiting for the girl. The girl is looking down at her hands. She is holding something small. She looks distracted. It looks like she's texting! *The Expected One* was painted by Ferdinand Georg Waldmüller in 1860. That is nearly 150 years before the first smartphone was invented! How do you explain that?

The Expected One was painted by Ferdinand Georg Waldmüller. He was also a writer.

Today, we have Photoshop to edit an image. Back then, artists didn't even have computers! But artists were still able to edit their painting. All they did was paint over it! This technique is called pentimento.

DEBUNKED

This painting definitely is evidence that time travel is real. Right? Unfortunately, no.

Again, people project information of what they know to fill in the gaps. The girl is holding something small in her hands. She does look distracted and like she's texting. But she isn't. The so-called smartphone she's holding is a small prayer book. These books were popular during that time. People would carry them around. According to art historians, the girl is actually on her way to church!

CHAPTER 5:

ANCIENT ASTRONAUT

CONSPIRACY

Since early human history, we've thought about space. We've looked to the stars with wonder. The study of space and stars is called astronomy. It has been around since ancient times. The earliest recorded observation of space was in 1600 BCE. It is a drawing of the Sun, Moon, and stars. But the first time we've ever been to space was in 1961. That year, Russian Yuri Gagarin became the first person to travel to space. People who travel to space are called astronauts.

Ancient astronauts are also called "ancient aliens." Pictured is a petroglyph from Italy. Petroglyphs are prehistoric drawings or carvings on rock.

Having knowledge of the past is called retrocognition.
Having knowledge of the future is called precognition.

In Salamanca, Spain, there is an old cathedral. A cathedral is a large, important church. This cathedral was built in the early 1500s. Sculptures are carved in its outside walls. One of the sculptures is of an astronaut! But astronauts didn't exist until the 1960s. Does this mean the sculptor traveled through time?

DEBUNKED

It's true we've been studying space since ancient history. We traveled to space in the 1960s. This is also true. There is an old cathedral in Salamanca. One of its walls does have a sculpture of an astronaut.

But there were renovations done to the cathedral. Renovations are updates. The astronaut sculpture was added during a renovation in 1992.

SPOT-LIGHT

SOCRATIC METHOD

The Socratic method is a learning tool. It involves asking questions until you find the truth. It is named after the ancient Greek philosopher Socrates. Philosophers study ideas like right and wrong. Socrates used this method on his students. The students would have a certain idea about something. Socrates would ask questions about that idea. The questions helped identify any gaps in the student's idea. The Socratic method is still used today. It is popular in law schools.

Let's take a look at the Socratic method in action!

Levi and Ari wonder if time travel is possible. Levi asks Ari, "Do we have any evidence of time travel?"

Ari answers, "Yes, there are old photos of people with phones."

Levi asks, "Could the phones be other objects?"

Ari says, "That is possible."

Levi asks, "What else could they be?"

Ari replies, "They could be books or small cases."

Levi asks, "Is this more likely?"

Ari says, "Yes."

Levi and Ari came closer to the truth through asking questions. The Socratic method can help disprove conspiracies!

CHAPTER 6:

THE VERDICT

Time travel is certainly a very curious subject. Imagine all the possibilities if it were real. Many odd things make it seem like it is. Where did a man in 1941 find modern clothing? What is a smartphone doing in an 1860 painting? Do astronauts travel back in time?

The answers, we now know, are quite simple. The man on the bridge is actually wearing the clothing of that time. The girl in the painting is praying, not texting! And astronauts only travel to space, not through time.

It seems these people were not time travelers after all. Moving beyond the "present" dot is impossible. The evidence supports this conclusion. This conspiracy theory has been DEBUNKED. Or has it? What do you think?

TRY THIS!

1. According to NASA, time portals exist! But not in the way we think. Research the Bermuda Triangle, the Philadelphia Experiment, and the Montauk Project.

2. Take a picture of yourself holding a small, flat object. Make it seem like it is a smartphone. Show the picture to a friend. Ask them what they think you are holding. Did you fool them?

3. Imagine you traveled back in time. Imagine you traveled to the time before smartphones existed. How would you describe the present time to someone? How would you describe social media?

4. Research Stephen Hawking. He was a famous scientist who believed in time travel. But not in the way we think. What were his thoughts on time travel? Discuss with a friend.

LEARN MORE

Goldstein, Margaret J. *What Are Conspiracy Theories?* Minneapolis, MN: Lerner Publications, 2020.

Johnson, Anna Maria. *Debunking Conspiracy Theories.* New York, NY: Cavendish Square Publishing, LLC, 2019.

GLOSSARY

biased (BY-uhst) favoring one idea because of personal preference

cathedral (kuh-THEE-druhl) a large, important church

confirmation bias (kahn-fuhr-MAY-shuhn BYE-uhs) noticing only information that supports one's view

conspiracies (kuhn-SPIHR-uh-seez) secret plans to do something bad or against the law

credible (CREH-duh-buhl) trusted or able to be proven

emblem (EHM-bluhm) a recognizable symbol

evidence (EH-vuh-duhnss) facts or information that supports a claim

motivated reasoning (MOH-tuh-vay-tuhd REEZ-ning) accepting information we already agree with

mysteries (MIH-stuh-reez) things that are unknown or hard to explain

perspective (puhr-SPEK-tiv) point of view

philosopher (fuh-LAH-suh-fuhr) person who studies questions and ideas about knowledge and right and wrong

project (pruh-JEKT) to assign one's own ideas to others

relativity (reh-luh-TIH-vuh-tee) theory that the speed of time is relative

renovations (reh-nuh-VAY-shuhns) updates

science fiction (SYE-uhns FIK-shuhn) imaginary stories about space, technology, or science

Socratic method (suh-KRAH-tik MEH-thuhd) asking questions until you find the truth

theorists (THEE-uh-rists) people who explain things with ideas called theories

time dilation (TYME dye-LAY-shuhn) the idea that the faster you move, the slower time passes

INDEX